FUNNEL WEB SPIDERS

THE SPIDER DISCOVERY LIBRARY

Louise Martin

Rourke Enterprises, Inc.
Vero Beach, Florida 32964

LIBRARY OF CONGRESS
Library of Congress Cataloging-in-Publication Data

Martin, Louise, 1955-
  Funnel web spiders / by Louise Martin.

  p. cm. — (The Spider discovery library)
  Includes index.
  Summary: Describes the physical characteristics,
habits, and natural environment of the Australian spider
which gets its name from the shape of its web and is one of
the most poisonous spiders in the world.
  ISBN 0-86592-962-9
  1. Agelenidae — Juvenile literature. [1. Funnel - web
spiders. 2. Spiders.] I. Title. II. Series:
Martin, Louise, 1955-
Spider discovery library.
QL458.42A3M37   1988
595.4'4 - dc19   88-5977
                                   CIP
                                   AC

Printed in the USA

*Title page photo: A tree funnel web
spider*

TABLE OF CONTENTS

FUNNEL WEB SPIDERS

Funnel web spiders take their name from the shape of their webs. The spiders spin a funnel-shaped web of silk at the entrance to their burrows. Funnel web spiders belong to the *Mygalomorph* group of spiders. They are dull black in color. Like other *Mygalomorphs*, funnel web spiders have hairy bodies and legs.

A funnel web spider in its web

WHERE THEY LIVE

Funnel web spiders are the most deadly spiders in the world. People are very frightened of being bitten by this kind of spider. There are no funnel web spiders in the United States. They live only in Australia. These spiders are often called Sydney funnel webs, after the large city of Sydney, Australia. Like many other spiders in the *Mygalomorph* group, funnel web spiders live underground. They spin tubes of silk in holes in the ground, and spend most of their time there in their cozy burrows. Only the silky, funnel-shaped web at the mouth of their burrows gives any clue as to where the funnel web spiders' homes may be.

A deadly Sydney funnel web spider

THEIR BURROWS

Compared to some of the other *Mygalomorph* spiders, funnel web spiders are not large. Females are normally about one and one-half inches across. They are shy and spend most of their lives in their burrows. Female funnel web spiders are deadly even to their own kind. They often eat male spiders.

A funnel web spider approaches its burrow

PREY

Funnel web spiders usually feed on insects, which they catch close to their burrows. The female spiders sit just inside the entrance to their burrows, waiting for passing insects. They are always ready to jump out and eat anything that sounds as though it would make a tasty meal. Funnel web spiders do not use their webs for catching their **prey**.

*A funnel web spider from
Queensland, Australia*

Funnel web spiders use their sense
of feeling more than their eyes

Funnel web spiders are the most poisonous in the world

THEIR SENSES

Although funnel web spiders have eight eyes, like other *Mygalomorph* spiders, their eyesight is poor. Funnel web spiders use their sense of feeling more than their eyes for hunting and catching their prey. The movement of insects passing their burrows is picked up by tiny hairs all over their bodies. The hairs **vibrate** and send messages to the spiders' brains, telling them that there is something close by.

This Australian spider is related to the deadly funnel web spiders

THEIR DEFENSES

When they are frightened, funnel web spiders do two things. First they lift their front legs to expose their poison fangs, and then they often bite. Sometimes female funnel web spiders will squeeze out a tiny drop of poison onto the tip of their fangs when in the defensive **posture**. This is to let their enemy know that their threats are serious!

Funnel web spiders' webs are very complex

FUNNEL WEBS AND PEOPLE

Luckily for the people who live in Australia, funnel web spiders do not come inside houses. Sometimes though they fall into people's swimming pools. They cannot move in the water and look as though they are dead. But once they are out of water, the spiders quickly wake up. If they think that they are being threatened, the deadly funnel web spiders will bite.

Spiders bite in self defense

FUNNEL WEB SPIDERS' BITES

When funnel web spiders bite people, their deadly **venom** can kill. The poison acts very quickly. The pain begins in the area of the bite, and spreads throughout the body, arms, and legs. The victim feels ill and feverish, and collapses. He or she finds it difficult to breathe and might turn blue due to lack of oxygen. He or she may even begin to froth at the mouth. The victim must be taken to a hospital immediately.

People are very frightened by funnel web spiders

TREATING VICTIMS

Until recently, there was no cure for those bitten by funnel web spiders. Several people died from these bites. If a victim was already weak, he or she had little chance for survival. It was very frightening to see a funnel web spider. Today people no longer die from funnel web spider bites. Fortunately scientists have developed an **antitoxin** to help victims.

Glossary

antitoxin (ANT i TOX in) — a drug that fights poison

posture (POS ture) — body position

prey (PREY) — an animal that is hunted for food

venom (VEN om) — poison

vibrate (vi BRATE) — to shake

INDEX